The Congregation for the Doctrine of the Faith

DONUM VITAE

Instruction on Respect for Human Life in Its
Origin and on the Dignity of Procreation

Replies to Certain Questions of the Day

February 22, 1987

En Route Books and Media, LLC
Saint Louis, MO

⊕*ENROUTE*
Make the time

En Route Books and Media, LLC

5705 Rhodes Avenue

St. Louis, MO 63109

Contact us at contactus@enroutebooksandmedia.com

Cover Design: Sebastian Mahfood using Dall-E for front image design with front and back cover image of Molly at 5 days after conception and at 1 year old.

ISBN: 979-8-88870-380-9

CONTENTS

i

FOREWORD

The Congregation for the Doctrine of the Faith has been approached by various Episcopal Conferences or individual Bishops, by theologians, doctors and scientists, concerning biomedical techniques which make it possible to intervene in the initial phase of the life of a human being and in the very processes of procreation and their conformity with the principles of Catholic morality. The present Instruction, which is the result of wide consultation and in particular of a careful evaluation of the declarations made by Episcopates, does not intend to repeat all the Church's teaching on the dignity of human life as it originates and on procreation, but to offer, in the light of the previous teaching of the Magisterium, some specific replies to the main questions being asked in this regard. The exposition is arranged as follows: an introduction will recall the fundamental principles, of an anthropological and moral character, which are necessary for a proper evaluation of the problems and for working out replies to those questions; the first part will have as its subject respect for the human being from the first moment of his or her existence; the second part will deal with the moral questions raised by technical interventions on human procreation; the third part will offer some orientations on the relationships between moral law and civil law in terms of the respect due to human embryos and foetuses[1] and as regards the legitimacy of techniques of artificial procreation.

---

[1] The terms "zygote", "pre-embryo", "embryo" and "foetus" can indicate in the vocabulary of biology successive stages of the development of a human being. The present Instruction makes free use of these terms, attributing to them an identical ethical relevance, in order to designate the result

(whether visible or not) of human generation, from the first moment of its existence until birth. The reason for this usage is clarified by the text (cf. I, 1).

INTRODUCTION

1. Biomedical Research and the Teaching of the Church

The gift of life which God the Creator and Father has entrusted to man calls him to appreciate the inestimable value of what he has been given and to take responsibility for it: this fundamental principle must be placed at the centre of one's reflection in order to clarify and solve the moral problems raised by artificial interventions on life as it originates and on the processes of procreation. Thanks to the progress of the biological and medical sciences, man has at his disposal ever more effective therapeutic resources; but he can also acquire new powers, with unforeseeable consequences, over human life at its very beginning and in its first stages. Various procedures now make it possible to intervene not only in order to assist but also to dominate the processes of procreation.

These techniques can enable man to "take in hand his own destiny", but they also expose him "to the temptation to go beyond the limits of a reasonable dominion over nature".(1) They might constitute progress in the service of man, but they also involve serious risks. Many people are therefore expressing an urgent appeal that in interventions on procreation the values and rights of the human person be safeguarded. Requests for clarification and guidance are coming not only from the faithful but also from those who recognize the Church as "an expert in humanity " (2) with a mission to serve the "civilization of love" (3) and of life.

The Church's Magisterium does not intervene on the basis of a particular competence in the area of the experimental sciences; but having taken account of the data of research and technology, it intends to put forward, by virtue of its evangelical mission and apostolic duty, the moral teaching corresponding to the dignity of the person and to his or her integral vocation. It intends to do so by expounding the criteria of moral judgment as regards the applications of scientific research and technology, especially in relation to human life and its beginnings. These criteria are the respect, defence and promotion of man, his "primary and fundamental right" to life,(4) his dignity as a person who is endowed with a spiritual soul and with moral responsibility (5) and who is called to beatific communion with God. The Church's intervention in this field is inspired also by the Love which she owes to man, helping him to recognize and respect his rights and duties. This love draws from the fount of Christ's love: as she contemplates the mystery of the Incarnate Word, the Church also comes to understand the "mystery of man"; (6) by proclaiming the Gospel of salvation, she reveals to man his dignity and invites him to discover fully the truth of his own being. Thus, the Church once more puts forward the divine law in order to accomplish the work of truth and liberation. For it is out of goodness - in order to indicate the path of life - that God gives human beings his commandments and the grace to observe them: and it is likewise out of goodness - in order to help them persevere along the same path - that God always offers to everyone his forgiveness. Christ has compassion on our weaknesses: he is our Creator and Redeemer. May his spirit open men's hearts to the gift of God's peace and to an understanding of his precepts.

2. Science and Technology at the Service of the Human Person

God created man in his own image and likeness: "male and female he created them" (*Gen* 1: 27), entrusting to them the task of "having dominion over the earth" (*Gen* 1:28). Basic scientific research and applied research constitute a significant expression of this dominion of man over creation. Science and technology are valuable resources for man when placed at his service and when they promote his integral development for the benefit of all; but they cannot of themselves show the meaning of existence and of human progress. Being ordered to man, who initiates and develops them, they draw from the person and his moral values the indication of their purpose and the awareness of their limits.

It would on the one hand be illusory to claim that scientific research and its applications are morally neutral; on the other hand, one cannot derive criteria for guidance from mere technical efficiency, from research's possible usefulness to some at the expense of others, or, worse still, from prevailing ideologies. Thus science and technology require, for their own intrinsic meaning, an unconditional respect for the fundamental criteria of the moral law: that is to say, they must be at the service of the human person, of his inalienable rights and his true and integral good according to the design and will of God.(7) The rapid development of technological discoveries gives greater urgency to this need to respect the criteria just mentioned: science without conscience can only lead to man's ruin. "Our era needs such wisdom more than bygone ages if the discoveries made by man are to be further humanized. For the future of the world stands in peril unless wiser people are forthcoming".(8)

3. Anthropology and Procedures in the Biomedical Field

Which moral criteria must be applied in order to clarify the problems posed today in the field of biomedicine? The answer to this question presupposes a proper idea of the nature of the human person in his bodily dimension.

For it is only in keeping with his true nature that the human person can achieve self-realization as a "unified totality":(9) and this nature is at the same time corporal and spiritual. By virtue of its substantial union with a spiritual soul, the human body cannot be considered as a mere complex of tissues, organs and functions, nor can it be evaluated in the same way as the body of animals; rather it is a constitutive part of the person who manifests and expresses himself through it. The natural moral law expresses and lays down the purposes, rights and duties which are based upon the bodily and spiritual nature of the human person. Therefore this law cannot be thought of as simply a set of norms on the biological level; rather it must be defined as the rational order whereby man is called by the Creator to direct and regulate his life and actions and in particular to make use of his own body.(10) A first consequence can be deduced from these principles: an intervention on the human body affects not only the tissues, the organs and their functions but also involves the person himself on different levels. It involves, therefore, perhaps in an implicit but nonetheless real way, a moral significance and responsibility. Pope John Paul II forcefully reaffirmed this to the World Medical Association when he said: "Each human person, in his absolutely unique singularity, is constituted not only by his spirit, but by his body as well. Thus, in the body and through the body, one touches the person himself in his concrete reality. To respect the dignity of man consequently amounts to safeguarding this identity of the man *'corpore et anima unus'*, as the Second Vatican Council says (*Gaudium et Spes*, 14, par.1). It is on the basis of this anthropological vision that one is to find the fundamental criteria for decision-making in the case of procedures which are not strictly therapeutic, as, for example, those aimed at the improvement of the human biological condition".(11)

Applied biology and medicine work together for the integral good of human life when they come to the aid of a person stricken by illness and infirmity and when they respect his or her dignity as a creature of God. No biologist or doctor can reasonably claim, by virtue of his scientific competence, to be able to decide on people's origin and destiny. This norm must be applied in a particular way in the field of sexuality and procreation, in which man and woman actualize the fundamental values of love and life. God, who is love and life, has inscribed in man and woman the vocation to share in a special way in his mystery of personal communion and in his work as Creator and Father.(12) For this reason marriage possesses specific goods and values in its union and in procreation which cannot be likened to those existing in lower forms of life. Such values and meanings are of the personal order and determine from the moral point of view the meaning and limits of artificial interventions on procreation and on the origin of human life. These interventions are not to be rejected on the grounds that they are artificial. As such, they bear witness to the possibilities of the art of medicine. But they must be given a moral evaluation in reference to the dignity of the human person, who is called to realize his vocation from God to the gift of love and the gift of life.

4. Fundamental Criteria for a Moral Judgment

The fundamental values connected with the techniques of artificial human procreation are two: the life of the human being called into existence and the special nature of the transmission of human life in marriage. The moral judgment on such methods of artificial procreation must therefore be formulated in reference to these values.

Physical life, with which the course of human life in the world begins, certainly does not itself contain the whole of a person's value, nor does it represent the supreme good of man who is called to eternal life. However it does constitute in a certain way the "fundamental " value of life, precisely because upon this physical life all the other values of the person are based and developed.(13) The inviolability of the innocent human being's right to life "from the moment of conception until death" (14) is a sign and requirement of the very inviolability of the person to whom the Creator has given the gift of life. By comparison with the transmission of other forms of life in the universe, the transmission of human life has a special character of its own, which derives from the special nature of the human person. "The transmission of human life is entrusted by nature to a personal and conscious act and as such is subject to the all-holy laws of God: immutable and inviolable laws which must be recognized and observed. For this reason, one cannot use means and follow methods which could be licit in the transmission of the life of plants and animals" (15)

Advances in technology have now made it possible to procreate apart from sexual relations through the meeting *in vitro* of the germ-cells previously taken from the man and the woman. But what is technically possible is not for that very reason morally admissible. Rational reflection on the fundamental values of life and of human procreation is therefore indispensable for formulating a moral evaluation of such technological interventions on a human being from the first stages of his development.

5. Teachings of the Magisterium

On its part, the Magisterium of the Church offers to human reason in this field too the light of Revelation: the doctrine concerning man taught by the Magisterium contains many elements which throw light on the problems being faced here. From the moment of conception, the life of every human being is to be respected in an absolute way because man is the only creature on earth that God has "wished for himself " (16) and the spiritual soul of each man is "immediately created" by God; (17) his whole being bears the image of the Creator. Human life is sacred because from its beginning it involves "the creative action of God" (18) and it remains forever in a special relationship with the Creator, who is its sole end.(19) God alone is the Lord of life from its beginning until its end: no one can, in any circumstance, claim for himself the right to destroy directly an innocent human being. (20) Human procreation requires on the part of the spouses responsible collaboration with the fruitful love of

God; (21) the gift of human life must be actualized in marriage through the specific and exclusive acts of husband and wife, in accordance with the laws inscribed in their persons and in their union.(22)

I. Respect for Human Embryos

Careful reflection on this teaching of the Magisterium and on the evidence of reason, as mentioned above, enables us to respond to the numerous moral problems posed by technical interventions upon the human being in the first phases of his life and upon the processes of his conception.

1. What Respect is Due to the Human Embryo, Taking into Account His Nature and Identity?

The human being must be respected - as a person - from the very first instant of his existence.

The implementation of procedures of artificial fertilization has made possible various interventions upon embryos and human foetuses. The aims pursued are of various kinds: diagnostic and therapeutic, scientific and commercial. From all of this, serious problems arise. Can one speak of a right to experimentation upon human embryos for the purpose of scientific research? What norms or laws should be worked out with regard to this matter? The response to these problems presupposes a detailed reflection on the nature and specific identity - the word "status" is used - of the human embryo itself.

At the Second Vatican Council, the Church for her part presented once again to modern man her constant and certain doctrine according to which: "Life once conceived, must be protected with the utmost care; abortion and infanticide are abominable crimes". (23) More recently, the Charter of the Rights of the Family, published by

the Holy See, confirmed that "Human life must be absolutely respected and protected from the moment of conception".(24)

This Congregation is aware of the current debates concerning the
beginning of human life, concerning the individuality of the human
being and concerning the identity of the human person. The Congregation recalls the teachings found in the *Declaration on Procured
Abortion*:

> "From the time that the ovum is fertilized, a new life is begun
> which is neither that of the father nor of the mother; it is rather
> the life of a new human being with his own growth. It would
> never be made human if it were not human already. To this per-
> petual evidence ... modern genetic science brings valuable con-
> firmation. It has demonstrated that, from the first instant, the
> programme is fixed as to what this living being will be: a man,
> this individual man with his characteristic aspects already well
> determined. Right from fertilization is begun the adventure of a
> human life, and each of its great capacities requires time ... to find
> its place and to be in a position to act". (25)

This teaching remains valid and is further confirmed, if confirmation were needed, by recent findings of human biological science
which recognize that in the zygote[2] resulting from fertilization the
biological identity of a new human individual is already constituted.

---

[2] The zygote is the cell produced when the nuclei of the two gametes
have fused. N.B. *"Zygotum est cellula orta a fusione duorum gametum."*
(AAS)

Certainly no experimental datum can be in itself sufficient to bring us to the recognition of a spiritual soul; nevertheless, the conclusions of science regarding the human embryo provide a valuable indication for discerning by the use of reason a personal presence at the moment of this first appearance of a human life: how could a human individual not be a human person? The Magisterium has not expressly committed itself to an affirmation of a philosophical nature, but it constantly reaffirms the moral condemnation of any kind of procured abortion. This teaching has not been changed and is unchangeable.(26)

Thus the fruit of human generation, from the first moment of its existence, that is to say from the moment the zygote has formed, demands the unconditional respect that is morally due to the human being in his bodily and spiritual totality. The human being is to be respected and treated as a person from the moment of conception; and therefore from that same moment his rights as a person must be recognized, among which in the first place is the inviolable right of every innocent human being to life.

This doctrinal reminder provides the fundamental criterion for the solution of the various problems posed by the development of the biomedical sciences in this field: since the embryo must be treated as a person, it must also be defended in its integrity, tended and cared for, to the extent possible, in the same way as any other human being as far as medical assistance is concerned.

2. Is Prenatal Diagnosis Morally Licit?

If prenatal diagnosis respects the life and integrity of the embryo and the human foetus and is directed towards its safeguarding or healing as an individual, then the answer is affirmative.

For prenatal diagnosis makes it possible to know the condition of the embryo and of the foetus when still in the mother's womb. It permits, or makes it possible to anticipate earlier and more effectively, certain therapeutic, medical or surgical procedures.

Such diagnosis is permissible, with the consent of the parents after they have been adequately informed, if the methods employed safeguard the life and integrity of the embryo and the mother, without subjecting them to disproportionate risks.(27) But this diagnosis is gravely opposed to the moral law when it is done with the thought of possibly inducing an abortion depending upon the results: a diagnosis which shows the existence of a malformation or a hereditary illness must not be the equivalent of a death-sentence. Thus a woman would be committing a gravely illicit act if she were to request such a diagnosis with the deliberate intention of having an abortion should the results confirm the existence of a malformation or abnormality. The spouse or relatives or anyone else would similarly be acting in a manner contrary to the moral law if they were to counsel or impose such a diagnostic procedure on the expectant mother with the same intention of possibly proceeding to an abortion. So too the specialist would be guilty of illicit collaboration if, in conducting the diagnosis and in communicating its results, he were deliberately to contribute to establishing or favouring a link between prenatal diagnosis and abortion.

In conclusion, any directive or programme of the civil and health authorities or of scientific organizations which in any way were to favour a link between prenatal diagnosis and abortion, or which were to go as far as directly to induce expectant mothers to submit to prenatal diagnosis planned for the purpose of eliminating foetuses which are affected by malformations or which are carriers of hereditary illness, is to be condemned as a violation of the unborn child's right to life and as an abuse of the prior rights and duties of the spouses.

3. Are Therapeutic Procedures Carried Out on the Human Embryo Licit?

As with all medical interventions on patients, *one must uphold as licit procedures carried out on the human embryo which respect the life and integrity of the embryo and do not involve disproportionate risks for it but are directed towards its healing, the improvement of its condition of health, or its individual survival.*

Whatever the type of medical, surgical or other therapy, the free and informed consent of the parents is required, according to the deontological rules followed in the case of children. The application of this moral principle may call for delicate and particular precautions in the case of embryonic or foetal life.

The legitimacy and criteria of such procedures have been clearly stated by Pope John Paul II:

"A strictly therapeutic intervention whose explicit objective is the healing of various maladies such as those stemming from chromosomal defects will, in principle, be considered desirable, provided it is directed to the true promotion of the personal well-being of the individual without doing harm to his integrity or worsening his conditions of life. Such an intervention would indeed fall within the logic of the Christian moral tradition" (28)

4. How Is One Morally to Evaluate Research and Experimentation[3] on Human Embryos and Foetuses?

Medical research must refrain from operations on live embryos, unless there is a moral certainty of not causing harm to the life or integrity of the unborn child and the mother, and on condition that the parents have givers their free and in formed consent to the procedure. It follows that all research, even when limited to the simple observation of the embryo, would become illicit were it to involve risk to the

---

[3] Since the terms "research" and "experimentation" are often used equivalently and ambiguously, it is deemed necessary to specify the exact meaning given them in this document.

1) By *research* is meant any inductive-deductive process which aims at promoting the systematic observation of a given phenomenon in the human field or at verifying a hypothesis arising from previous observations.

2) By *experimentation* is meant any research in which the human being (in the various stages of his existence: embryo, foetus, child or adult) represents the object through which or upon which one intends to verify the effect, at present unknown or not sufficiently known, of a given treatment (e.g., pharmacological, teratogenic, surgical, etc.).

embryo's physical integrity or life by reason of the methods used or the effects induced.

As regards experimentation, and presupposing the general distinction between experimentation for purposes which are not directly therapeutic and experimentation which is clearly therapeutic for the subject himself, in the case in point one must also distinguish between experimentation carried out on embryos which are still alive and experimentation carried out on embryos which are dead. *If the embryos are living, whether viable or not, they must be respected just like any other human person; experimentation on embryos which is not directly therapeutic is illicit.*(29)

No objective, even though noble in itself, such as a foreseeable advantage to science, to other human beings or to society, can in any way justify experimentation on living human embryos or foetuses, whether viable or not, either inside or outside the mother's womb. The informed consent ordinarily required clinical experimentation on adults cannot be granted by the parents, who may not freely dispose of the physical integrity or life of the unborn child. Moreover, experimentation on embryos and foetuses always involves risk, and indeed in most cases it involves the certain expectation of harm to their physical integrity or even their death.

To use human embryos or foetuses as the object or instrument of experimentation constitutes a crime against their dignity as human beings having a right to the same respect that is due to the child already born and to every human person.

The *Charter of the Rights of the Family* published by the Holy See affirms: "Respect for the dignity of the human being excludes all experimental manipulation or exploitation of the human embryo".(30)

The practice of keeping alive human embryos *in vivo* or *in vitro* for experimental or commercial purposes is totally opposed to human dignity.

In the case of experimentation that is clearly therapeutic, namely, when it is a matter of experimental forms of therapy used for the benefit of the embryo itself in a final attempt to save its life, and in the absence of other reliable forms of therapy, recourse to drugs or procedures not yet fully tested can be licit (31)

The corpses of human embryos and foetuses, whether they have been deliberately aborted or not, must be respected just as the remains of other human beings. In particular, they cannot be subjected to mutilation or to autopsies if their death has not yet been verified and without the consent of the parents or of the mother. Furthermore, the moral requirements must be safeguarded that there be no complicity in deliberate abortion and that the risk of scandal be avoided. Also, in the case of dead foetuses, as for the corpses of adult persons, all commercial trafficking must be considered illicit and should be prohibited.

5. How is One Morally to Evaluate the Use for Research Purposes of Embryos Obtained by 'In Vitro' Fertilization?

Human embryos obtained *in vitro* are human beings and subjects with rights: their dignity and right to life must be respected from the first moment of their existence. *It is immoral to produce human embryos destined to be exploited as disposable "biological material".*

In the usual practice of *in vitro* fertilization, not all of the embryos are transferred to the woman's body; some are destroyed. Just as the Church condemns induced abortion, so she also forbids acts against the life of these human beings. *It is a duty to condemn the particular gravity of the voluntary destruction of human embryos obtained 'in vitro' for the sole purpose of research, either by means of artificial insemination or by means of "twin fission".* By acting in this way the researcher usurps the place of God; and, even though he may be unaware of this, he sets himself up as the master of the destiny of others inasmuch as he arbitrarily chooses whom he will allow to live and whom he will send to death and kills defenceless human beings.

Methods of observation or experimentation which damage or impose grave and disproportionate risks upon embryos obtained *in vitro* are morally illicit for the same reasons. Every human being is to be respected for himself, and cannot be reduced in worth to a pure and simple instrument for the advantage of others. *It is therefore not in conformity with the moral law deliberately to expose to death human embryos obtained 'in vitro'.* In consequence of the fact that they have been produced *in vitro*, those embryos which art not transferred into the body of the mother and are called "spare" are exposed to an absurd fate, with no possibility of their being offered safe means of survival which can be licitly pursued.

6. What Judgment Should Be Made on Other Procedures of Manipulating Embryos Connected with the "Techniques of Human Reproduction"?

Techniques of fertilization *in vitro* can open the way to other forms of biological and genetic manipulation of human embryos, such as attempts or plans for fertilization between human and animal gametes and the gestation of human embryos in the uterus of animals, or the hypothesis or project of constructing artificial uteruses for the human embryo. *These procedures are contrary to the human dignity proper to the embryo, and at the same time they are contrary to the right of every person to be conceived and to be born within marriage and from marriage.*(32) *Also, attempts or hypotheses for obtaining a human being without any connection with sexuality through "twin fission," cloning or parthenogenesis are to be considered contrary to the moral law, since they are in opposition to the dignity both of human procreation and of the conjugal union.*

The freezing of embryos, even when carried out in order to preserve the life of an embryo - cryopreservation - *constitutes an offence against the respect due to human beings* by exposing them to grave risks of death or harm to their physical integrity and depriving them, at least temporarily, of maternal shelter and gestation, thus placing them in a situation in which further offences and manipulation are possible.

Certain attempts to influence chromosomic or genetic inheritance are not therapeutic but are aimed at producing human beings selected according to sex or other predetermined qualities. These manipulations are contrary to the personal dignity of the human being and his or her integrity and identity. Therefore, in no way can they be justified on the grounds of possible beneficial consequences for future humanity. (33) Every person must be respected for himself: in this consists the dignity and right of every human being from his or her beginning.

II. Interventions upon Human Procreation

By "artificial procreation" or "artificial fertilization" are understood here the different technical procedures directed towards obtaining a human conception in a manner other than the sexual union of man and woman. This Instruction deals with fertilization of an ovum in a test-tube (*in vitro* fertilization) and artificial insemination through transfer into the woman's genital tracts of previously collected sperm.

A preliminary point for the moral evaluation of such technical procedures is constituted by the consideration of the circumstances and consequences which those procedures involve in relation to the respect due the human embryo. Development of the practice of *in vitro* fertilization has required innumerable fertilizations and destructions of human embryos. Even today, the usual practice presupposes a hyperovulation on the part of the woman: a number of ova are withdrawn, fertilized and then cultivated *in vitro* for some days. Usually not all are transferred into the genital tracts of the woman; some embryos, generally called "spare ", are destroyed or frozen. On occasion, some of the implanted embryos are sacrificed for various eugenic, economic or psychological reasons. Such deliberate destruction of human beings or their utilization for different purposes to the detriment of their integrity and life is contrary to the doctrine on procured abortion already recalled.

The connection between *in vitro* fertilization and the voluntary destruction of human embryos occurs too often. This is significant:

through these procedures, with apparently contrary purposes, life and death are subjected to the decision of man, who thus sets himself up as the giver of life and death by decree. This dynamic of violence and domination may remain unnoticed by those very individuals who, in wishing to utilize this procedure, become subject to it themselves. The facts recorded and the cold logic which links them must be taken into consideration for a moral judgment on IVF and ET (*in vitro* fertilization and embryo transfer): the abortion-mentality which has made this procedure possible thus leads, whether one wants it or not, to man's domination over the life and death of his fellow human beings and can lead to a system of radical eugenics.

Nevertheless, such abuses do not exempt one from a further and thorough ethical study of the techniques of artificial procreation considered in themselves, abstracting as far as possible from the destruction of embryos produced *in vitro*.

The present Instruction will therefore take into consideration in the first place the problems posed by heterologous artificial fertilization (II, 1-3),[4]

---

[4] By the term *heterologous artificial fertilization or procreation*, the Instruction means techniques used to obtain a human conception artificially by the use of gametes coming from at least one donor other than the spouses who are joined in marriage. Such techniques can be of two types

    a)   *Heterologous IVF and ET*: the technique used to obtain a human conception through the meeting *in vitro* of gametes taken from at least one donor other than the two spouses joined in marriage.

    b)   *Heterologous artificial insemination*: the technique used to obtain a human conception through the transfer into the genital tracts of and subsequently those linked with homologous artificial fertilization (II, 4-6).[5]

Before formulating an ethical judgment on each of these procedures, the principles and values which determine the moral evaluation of each of them will be considered.

A. HETEROLOGOUS ARTIFICIAL FERTILIZATION

1. Why Must Human Procreation Take Place in Marriage?

Every human being is always to be accepted as a gift and blessing of God. However, from the moral point of view a truly responsible procreation vis-a-vis the unborn child must be the fruit of marriage. For human procreation has specific characteristics by virtue of the personal dignity of the parents and of the children: the procreation of a new person, whereby the man and the woman collaborate with the

---

the woman of the sperm previously collected from a donor other than the husband.

[5] *By artificial homologous fertilization or procreation,* the Instruction means the technique used to obtain a human conception using the gametes of the two spouses joined in marriage. Homologous artificial fertilization can be carried out by two different methods:

a) *Homologous IVF and ET*: the technique used to obtain a human conception through the meeting *in vitro* of the gametes of the spouses joined in marriage.

b) *Homologous artificial insemination*: the technique used to obtain a human conception through the transfer into the genital tracts of a married woman of the sperm previously collected from her husband.

power of the Creator, must be the fruit and the sign of the mutual self-giving of the spouses, of their love and of their fidelity.(34) *The fidelity of the spouses in the unity of marriage involves reciprocal respect of their right to become a father and a mother only through each other.*

The child has the right to be conceived, carried in the womb, brought into the world and brought up within marriage: it is through the secure and recognized relationship to his own parents that the child can discover his own identity and achieve his own proper human development.

The parents find in their child a confirmation and completion of their reciprocal self-giving: the child is the living image of their love, the permanent sign of their conjugal union, the living and indissoluble concrete expression of their paternity and maternity, (35) By reason of the vocation and social responsibilities of the person, the good of the children and of the parents contributes to the good of civil society; the vitality and stability of society require that children come into the world within a family and that the family be firmly based on marriage.

The tradition of the Church and anthropological reflection recognizes in marriage and in its indissoluble unity the only setting worthy of truly responsible procreation.

2. Does Heterologous Artificial Fertilization Conform to the Dignity of the Couple and to the Truth of Marriage?

Through IVF and ET and heterologous artificial insemination, human conception is achieved through the fusion of gametes of at least one donor other than the spouses who are united in marriage. *Heterologous artificial fertilization is contrary to the unity of marriage, to the dignity of the spouses, to the vocation proper to parents, and to the child's right to be conceived and brought into the world in marriage and from marriage.*(36) Respect for the unity of marriage and for conjugal fidelity demands that the child be conceived in marriage; the bond existing between husband and wife accords the spouses, in an objective and inalienable manner, the exclusive right to become father and mother solely through each other.(37) Recourse to the gametes of a third person, in order to have sperm or ovum available, constitutes a violation of the reciprocal commitment of the spouses and a grave lack in regard to that essential property of marriage which is its unity. Heterologous artificial fertilization violates the rights of the child; it deprives him of his filial relationship with his parental origins and can hinder the maturing of his personal identity. Furthermore, it offends the common vocation of the spouses who are called to fatherhood and motherhood: it objectively deprives conjugal fruitfulness of its unity and integrity; it brings about and manifests a rupture between genetic parenthood, gestational parenthood and responsibility for upbringing. Such damage to the personal relationships within the family has repercussions on civil society: what threatens the unity and stability of the family is a source of dissension, disorder and injustice in the whole of social life. *These reasons lead to a negative moral judgment concerning heterologous artificial fertilization: consequently fertilization of a married woman with the sperm of a donor different from her husband and fertilization with the husband's sperm of an ovum not coming from his wife are morally illicit. Furthermore, the artificial fertilization of a woman who is unmarried or a widow, whoever the donor may be, cannot be morally justified.*

The desire to have a child and the love between spouses who long to obviate a sterility which cannot be overcome in any other way constitute understandable motivations; but subjectively good intentions do not render heterologous artificial fertilization conformable to the objective and inalienable properties of marriage or respectful of the rights of the child and of the spouses.

3. Is "Surrogate"[6] Motherhood Morally Licit?

No, for the same reasons which lead one to reject heterologous artificial fertilization: for it is contrary to the unity of marriage and to the dignity of the procreation of the human person. Surrogate motherhood represents an objective failure to meet the obligations of maternal love, of conjugal fidelity and of responsible motherhood;

---

[6] By "surrogate mother" the Instruction means:

a) the woman who carries in pregnancy an embryo implanted in her uterus and who is genetically a stranger to the embryo because it has been obtained through the union of the gametes of "donors". She carries the pregnancy with a pledge to surrender the baby once it is born to the party who commissioned or made the agreement for the pregnancy.

b) the woman who carries in pregnancy an embryo to whose procreation she has contributed the donation of her own ovum, fertilized through insemination with the sperm of a man other than her husband. She carries the pregnancy with a pledge to surrender the child once it is born to the party who commissioned or made the agreement for the pregnancy.

it offends the dignity and the right of the child to be conceived, carried in the womb, brought into the world and brought up by his own parents; it sets up, to the detriment of families, a division between the physical, psychological and moral elements which constitute those families.

B. HOMOLOGOUS ARTIFICIAL FERTILIZATION

Since heterologous artificial fertilization has been declared unacceptable, the question arises of how to evaluate morally the process of homologous artificial fertilization: IVF and ET and artificial insemination between husband and wife. First a question of principle must be clarified.

4. What Connection Is Required from the Moral Point of View between Procreation and the Conjugal Act?

a) The Church's teaching on marriage and human procreation affirms the "inseparable connection, willed by God and unable to be broken by man on his own initiative, between the two meanings of the conjugal act: the unitive meaning and the procreative meaning. Indeed, by its intimate structure, the conjugal act, while most closely uniting husband and wife, capacitates them for the generation of new lives, according to laws inscribed in the very being of man and of woman".(38) This principle, which is based upon the nature of marriage and the intimate connection of the goods of marriage, has well-known consequences on the level of responsible fatherhood and motherhood. "By safeguarding both these essential aspects, the unitive and the procreative, the conjugal act preserves in its fullness the sense of true mutual love and its ordination towards man's exalted vocation to parenthood".(39)

The same doctrine concerning the link between the meanings of the conjugal act and between the goods of marriage throws light on the moral problem of homologous artificial fertilization, since "it is never permitted to separate these different aspects to such a degree as positively to exclude either the procreative intention or the conjugal relation" (40)

Contraception deliberately deprives the conjugal act of its openness to procreation and in this way brings about a voluntary dissociation of the ends of marriage. Homologous artificial fertilization, in seeking a procreation which is not the fruit of a specific act of conjugal union, objectively effects an analogous separation between the goods and the meanings of marriage. Thus, *fertilization is licitly sought when it is the result of a "conjugal act which is per se suitable for the generation of children to which marriage is ordered by its nature and by which the spouses become one flesh".(41) But from the moral point of view procreation is deprived of its proper perfection when it is not desired as the fruit of the conjugal act, that is to say of the specific act of the spouses' union.*

b) The moral value of the intimate link between the goods of marriage and between the meanings of the conjugal act is based upon the unity of the human being, a unity involving body and spiritual soul. (42) Spouses mutually express their personal love in the "language of the body ", which clearly involves both "spousal meanings" and parental ones.(43) The conjugal act by which the couple mutually express their self-gift at the same time expresses openness to the gift of life. It is an act that is inseparably corporal and spiritual. It is in their bodies and through their bodies that the spouses consummate their marriage and are able to become father and mother. In order to respect the language of their bodies and their natural generosity, the conjugal union must take place with respect for its openness to procreation; and the procreation of a person must be the fruit and the result of married love. The origin of the human being thus follows from a procreation that is "linked to the union, not only biological but also spiritual, of the parents, made one by the bond of marriage".(44) Fertilization achieved outside the bodies of the couple remains by this very fact deprived of the meanings and the values which are expressed in the language of the body and in the union of human persons.

c) Only respect for the link between the meanings of the conjugal act and respect for the unity of the human being make possible procreation in conformity with the dignity of the person. In his unique and irrepeatable origin, the child must be respected and recognized as equal in personal dignity to those who give him life. The human person must be accepted in his parents' act of union and love; the generation of a child must therefore be the fruit of that mutual giving (45) which is realized in the conjugal act wherein the spouses cooperate as servants and not as masters in the work of the Creator who is Love. In reality, the origin of a human person is the result of an act of giving. The one conceived must be the fruit of his parents' love. He cannot be desired or conceived as the product of an intervention of medical or biological techniques; that would be equivalent to reducing him to an object of scientific technology. No one may subject the coming of a child into the world to conditions of technical efficiency which are to be evaluated according to standards of control and dominion. *The moral relevance of the link between the meanings of the conjugal act and between the goods of marriage, as well as the unity of the human being and the dignity of his origin, demand that the procreation of a human person be brought about as the fruit of the conjugal act specific to the love between spouses.*(46) The link between procreation and the conjugal act is thus shown to be of great importance on the anthropological and moral planes, and it throws light on the positions of the Magisterium with regard to homologous artificial fertilization.

5. Is Homologous 'In Vitro' Fertilization Licit?

The answer to this question is strictly dependent on the principles just mentioned. Certainly one cannot ignore the legitimate aspirations of sterile couples. For some, recourse to homologous IVF and ET appears to be the only way of fulfilling their sincere desire for a child. The question is asked whether the totality of conjugal life in such situations is not sufficient to ensure the dignity proper to human procreation. It is acknowledged that IVF and ET certainly cannot supply for the absence of sexual relations (47) and cannot be preferred to the specific acts of conjugal union, given the risks involved for the child and the difficulties of the procedure. But it is asked whether, when there is no other way of overcoming the sterility which is a source of suffering, homologous *in vitro* fertilization may not constitute an aid, if not a form of therapy, whereby its moral licitness could be admitted.

The desire for a child - or at the very least an openness to the transmission of life - is a necessary prerequisite from the moral point of view for responsible human procreation. But this good intention is not sufficient for making a positive moral evaluation of *in vitro* fertilization between spouses. The process of IVF and ET must be judged in itself and cannot borrow its definitive moral quality from the totality of conjugal life of which it becomes part nor from the conjugal acts which may precede or follow it.(48)

It has already been recalled that, in the circumstances in which it is regularly practised, IVF and ET involves the destruction of human beings, which is something contrary to the doctrine on the illicitness of abortion previously mentioned.(49) But even in a situation in which every precaution were taken to avoid the death of human embryos, homologous IVF and ET dissociates from the conjugal act the actions which are directed to human fertilization. For this reason the very nature of homologous IVF and ET also must be taken into account, even abstracting from the link with procured abortion. Homologous IVF and ET is brought about outside the bodies of the couple through actions of third parties whose competence and technical activity determine the success of the procedure. Such fertilization entrusts the life and identity of the embryo into the power of doctors and biologists and establishes the domination of technology over the origin and destiny of the human person. Such a relationship of domination is in itself contrary to the dignity and equality that must be common to parents and children.

Conception *in vitro* is the result of the technical action which presides over fertilization. *Such fertilization is neither in fact achieved nor positively willed as the expression and fruit of a specific act of the conjugal union. In homologous IVF and ET, therefore, even if it is considered in the context of 'de facto' existing sexual relations, the generation of the human person is objectively deprived of its proper perfection: namely, that of being the result and fruit of a conjugal act* in which the spouses can become "cooperators with God for giving life to a new person".(50)

These reasons enable us to understand why the act of conjugal love is considered in the teaching of the Church as the only setting worthy of human procreation. For the same reasons the so-called "simple case," i.e., a homologous IVF and ET procedure that is free of any compromise with the abortive practice of destroying embryos and with masturbation, remains a technique which is morally illicit because it deprives human procreation of the dignity which is proper and connatural to it.

Certainly, homologous IVF and ET fertilization is not marked by all that ethical negativity found in extra-conjugal procreation; the family and marriage continue to constitute the setting for the birth and upbringing of the children. Nevertheless, in conformity with the traditional doctrine relating to the goods of marriage and the dignity of the person, *the Church remain opposed from the moral point of view to homologous 'in vitro' fertilization. Such fertilization is in itself illicit and in opposition to the dignity of procreation and of the conjugal union, even when everything is done to avoid the death of the human embryo.*

Although the manner in which human conception is achieved with IVF and ET cannot be approved, every child which comes into the world must in any case be accepted as a living gift of the divine Goodness and must be brought up with love.

6. How is Homologous Artificial Insemination to Be Evaluated from the Moral Point of View?

Homologous artificial insemination within marriage cannot be admitted except for those cases in which the technical means is not a substitute for the conjugal act but serves to facilitate and to help so that the act attains its natural purpose.

The teaching of the Magisterium on this point has already been stated.(51) This teaching is not just an expression of particular historical circumstances but is based on the Church's doctrine concerning the connection between the conjugal union and procreation and on a consideration of the personal nature of the conjugal act and of human procreation. "In its natural structure, the conjugal act is a personal action, a simultaneous and immediate cooperation on the part of the husband and wife, which by the very nature of the agents and the proper nature of the act is the expression of the mutual gift which, according to the words of Scripture, brings about union 'in one flesh' ".(52) Thus moral conscience "does not necessarily proscribe the use of certain artificial means destined solely either to the facilitating of the natural act or to ensuring that the natural act normally performed achieves its proper end".(53) If the technical means facilitates the conjugal act or helps it to reach its natural objectives, it can be morally acceptable. If, on the other hand, the procedure were to replace the conjugal act, it is morally illicit.

Artificial insemination as a substitute for the conjugal act is prohibited by reason of the voluntarily achieved dissociation of the two meanings of the conjugal act. Masturbation, through which the sperm is normally obtained, is another sign of this dissociation: even when it is done for the purpose of procreation, the act remains deprived of its unitive meaning: "It lacks the sexual relationship called for by the moral order, namely the relationship which realizes 'the full sense of mutual self-giving and human procreation in the context of true love' ". (54)

7. What Moral Criterion Can Be Proposed with Regard to Medical Intervention in Human Procreation?

The medical act must be evaluated not only with reference to its technical dimension but also and above all in relation to its goal which is the good of persons and their bodily and psychological health. The moral criteria for medical intervention in procreation are deduced from the dignity of human persons, of their sexuality and of their origin.

Medicine which seeks to be ordered to the integral good of the person must respect the specifically human values of sexuality.(55) *The doctor is at the service of persons and of human procreation. He does not have the authority to dispose of them or to decide their fate. A* medical intervention respects the dignity of persons when it seeks to assist the conjugal act either in order to facilitate its performance or in order to enable it to achieve its objective once it has been normally performed",(56)

On the other hand, it sometimes happens that a medical procedure technologically replaces the conjugal act in order to obtain a procreation which is neither its result nor its fruit. In this case the medical act is not, as it should be, at the service of conjugal union but rather appropriates to itself the procreative function and thus contradicts the dignity and the inalienable rights of the spouses and of the child to be born.

The humanization of medicine, which is insisted upon today by everyone, requires respect for the integral dignity of the human person first of all in the act and at the moment in which the spouses transmit life to a new person. It is only logical therefore to address an urgent appeal to Catholic doctors and scientists that they bear exemplary witness to the respect due to the human embryo and to the dignity of procreation. The medical and nursing staff of Catholic hospitals and clinics are in a special way urged to do justice to the moral obligations which they have assumed, frequently also, as part of their contract. Those who are in charge of Catholic hospitals and clinics and who are often Religious will take special care to safeguard and promote a diligent observance of the moral norms recalled in the present Instruction.

8. The Suffering Caused by Infertility in Marriage

The suffering of spouses who cannot have children or who are afraid of bringing a handicapped child into the world is a suffering that everyone must understand and properly evaluate.

On the part of the spouses, the desire for a child is natural: it expresses the vocation to fatherhood and motherhood inscribed in conjugal love. This desire can be even stronger if the couple is affected by sterility which appears incurable. Nevertheless, marriage does not confer upon the spouses the right to have a child, but only the right to perform those natural acts which are *per se* ordered to procreation.(57)

A true and proper right to a child would be contrary to the child's dignity and nature. The child is not an object to which one has a right, nor can he be considered as an object of ownership: rather, a child is a gift, "the supreme gift" (58) *and the most gratuitous gift of marriage, and is a living testimony of the mutual giving of his parents. For this reason, the child has the right, as already mentioned, to be the fruit of the specific act of the conjugal love of his parents; and he also has the right to be respected as a person from the moment of his conception.*

Nevertheless, whatever its cause or prognosis, sterility is certainly a difficult trial. The community of believers is called to shed light upon and support the suffering of those who are unable to fulfill their legitimate aspiration to motherhood and fatherhood. Spouses who find themselves in this sad situation are called to find in it an opportunity for sharing in a particular way in the Lord's Cross, the source of spiritual fruitfulness. Sterile couples must not forget that "even when procreation is not possible, conjugal life does not for this reason lose its value. Physical sterility in fact can be for spouses the occasion for other important services to the life of the human person, for example, adoption, various forms of educational work, and assistance to other families and to poor or handicapped children".(59)

Many researchers are engaged in the fight against sterility. While fully safeguarding the dignity of human procreation, some have achieved results which previously seemed unattainable. Scientists therefore are to be encouraged to continue their research with the aim of preventing the causes of sterility and of being able to remedy them so that sterile couples will be able to procreate in full respect for their own personal dignity and that of the child to be born.

III. MORAL AND CIVIL LAW

THE VALUES AND MORAL OBLIGATIONS THAT CIVIL LEGISLATION MUST RESPECT AND SANCTION IN THIS MATTER

The inviolable right to life of every innocent human individual and the rights of the family and of the institution of marriage constitute fundamental moral values, because they concern the natural condition and integral vocation of the human person; at the same time they are constitutive elements of civil society and its order.

For this reason the new technological possibilities which have opened in the field of biomedicine require the intervention of the political authorities and of the legislator, since an uncontrolled application of such techniques could lead to unforeseeable and damaging consequences for civil society. Recourse to the conscience of each individual and to the self-regulation of researchers cannot be sufficient for ensuring respect for personal rights and public order. If the legislator responsible for the common good were not watchful, he could be deprived of his prerogatives by researchers claiming to govern humanity in the name of the biological discoveries and the alleged "improvement" processes which they would draw from those discoveries. "Eugenism" and forms of discrimination between human beings could come to be legitimized: this would constitute an act of violence and a serious offense to the equality, dignity and fundamental rights of the human person.

The intervention of the public authority must be inspired by the rational principles which regulate the relationships between civil law and moral law. The task of the civil law is to ensure the common good of people through the recognition of and the defence of fundamental rights and through the promotion of peace and of public morality.(60) In no sphere of life can the civil law take the place of conscience or dictate norms concerning things which are outside its competence. It must sometimes tolerate, for the sake of public order, things which it cannot forbid without a greater evil resulting. However, the inalienable rights of the person must be recognized and respected by civil society and the political authority. These human rights depend neither on single individuals nor on parents; nor do they represent a concession made by society and the State: they pertain to human nature and are inherent in the person by virtue of the creative act from which the person took his of her origin.

Among such fundamental rights one should mention in this regard: a) every human being's right to life and physical integrity from the moment of conception until death; b) the rights of the family and of marriage as an institution and, in this area, the child's right to be conceived, brought into the world and brought up by his parents. To each of these two themes it is necessary here to give some further consideration.

In various States certain laws have authorized the direct suppression of innocents: the moment a positive law deprives a category of human beings of the protection which civil legislation must accord them, the State is denying the equality of all before the law. When the State does not place its power at the service of the rights of each citizen, and in particular of the more vulnerable, the very foundations of a State based on law are undermined. The political authority consequently cannot give approval to the calling of human beings into existence through procedures which would expose them to those very grave risks noted previously. The possible recognition by positive law and the political authorities of techniques of artificial transmission of life and the experimentation connected with it would widen the breach already opened by the legalization of abortion.

As a consequence of the respect and protection which must be ensured for the unborn child from the moment of his conception, the law must provide appropriate penal sanctions for every deliberate violation of the child's rights. The law cannot tolerate - indeed it must expressly forbid - that human beings, even at the embryonic stage, should be treated as objects of experimentation, be mutilated or destroyed with the excuse that they are superfluous or incapable of developing normally.

The political authority is bound to guarantee to the institution of the family, upon which society is based, the juridical protection to which it has a right. From the very fact that it is at the service of people, the political authority must also be at the service of the family.

Civil law cannot grant approval to techniques of artificial procreation which, for the benefit of third parties (doctors, biologists, economic or governmental powers), take away what is a right inherent in the relationship between spouses; and therefore civil law cannot legalize the donation of gametes between persons who are not legitimately united in marriage.

Legislation must also prohibit, by virtue of the support which is due to the family, embryo banks, *post mortem* insemination and "surrogate motherhood."

It is part of the duty of the public authority to ensure that the civil law is regulated according to the fundamental norms of the moral law in matters concerning human rights, human life and the institution of the family. Politicians must commit themselves, through their interventions upon public opinion, to securing in society the widest possible consensus on such essential points and to consolidating this consensus wherever it risks being weakened or is in danger of collapse.

In many countries, the legalization of abortion and juridical tolerance of unmarried couples makes it more difficult to secure respect for the fundamental rights recalled by this Instruction. It is to be hoped that States will not become responsible for aggravating these socially damaging situations of injustice. It is rather to be hoped that nations and States will realize all the cultural, ideological and political implications connected with the techniques of artificial procreation and will find the wisdom and courage necessary for issuing laws which are more just and more respectful of human life and the institution of the family.

The civil legislation of many states confers an undue legitimation upon certain practices in the eyes of many today; it is seen to be incapable of guaranteeing that morality which is in conformity with the natural exigencies of the human person and with the "unwritten laws" etched by the Creator upon the human heart. All men of good will must commit themselves, particularly within their professional field and in the exercise of their civil rights, to ensuring the reform of morally unacceptable civil laws and the correction of illicit practices. In addition, "conscientious objection" vis-a-vis such laws must be supported and recognized. A movement of passive resistance to the legitimation of practices contrary to human life and dignity is beginning to make an ever sharper impression upon the moral conscience of many, especially among specialists in the biomedical sciences.

Conclusion

The spread of technologies of intervention in the processes of human procreation raises very serious moral problems in relation to the respect due to the human being from the moment of conception, to the dignity of the person, of his or her sexuality, and of the transmission of life.

With this Instruction the Congregation for the Doctrine of the Faith, in fulfilling its responsibility to promote and defend the Church's teaching in so serious a matter, addresses a new and heartfelt invitation to all those who, by reason of their role and their commitment, can exercise a positive influence and ensure that, in the family and in society, due respect is accorded to life and love. It addresses this invitation to those responsible for the formation of consciences and of public opinion, to scientists and medical professionals, to jurists and politicians. It hopes that all will understand the incompatibility between recognition of the dignity of the human person and contempt for life and love, between faith in the living God and the claim to decide arbitrarily the origin and fate of a human being.

In particular, the Congregation for the Doctrine of the Faith addresses an invitation with confidence and encouragement to theologians, and above all to moralists, that they study more deeply and make ever more accessible to the faithful the contents of the teaching of the Church's Magisterium in the light of a valid anthropology in the matter of sexuality and marriage and in the context of the necessary interdisciplinary approach. Thus they will make it possible to understand ever more clearly the reasons for and the validity of this teaching. By defending man against the excesses of his own power, the Church of God reminds him of the reasons for his true nobility; only in this way can the possibility of living and loving with that dignity and liberty which derive from respect for the truth be ensured for the men and women of tomorrow. The precise indications which are offered in the present Instruction therefore are not meant to halt the effort of reflection but rather to give it a renewed impulse in unrenounceable fidelity to the teaching of the Church.

In the light of the truth about the gift of human life and in the light of the moral principles which flow from that truth, everyone is invited to act in the area of responsibility proper to each and, like the good Samaritan, to recognize as a neighbour even the littlest among the children of men (Cf. *Lk* 10: 2 9-37). Here Christ's words find a new and particular echo: "What you do to one of the least of my brethren, you do unto me" (*Mt* 25:40).

During an audience granted to the undersigned Prefect after the plenary session of the Congregation for the Doctrine of the Faith, the Supreme Pontiff, John Paul II, approved this Instruction and ordered it to be published.

Given at Rome, from the Congregation for the Doctrine of the Faith, February 22, 1987, the Feast of the Chair of St. Peter, the Apostle.

JOSEPH CARDINAL RATZINGER
Prefect

ALBERTO BOVONE
Titular Archbishop of Caesarea in Numidia
Secretary

Notes

(1) POPE JOHN PAUL II, *Discourse to those taking part in the 81st Congress of the Italian Society of Internal Medicine and the 82nd Congress of the Italian Society of General Surgery*, 27 October 1980: AAS 72 (1980) 1126.

(2) POPE PAUL VI, *Discourse to the General Assembly of the United Nations Organization*, 4 October 1965: AAS 57 (1965) 878; Encyclical *Populorum Progressio*, 13: AAS 59 (1967) 263.

(3) POPE PAUL VI, *Homily during the Mass closing the Holy Year*, 25 December 1975: AAS 68 (1976) 145; POPE JOHN PAUL II, Encyclical *Dives in Misericordia*, 30: AAS 72 (1980) 1224.

(4) POPE JOHN PAUL II, *Discourse to those taking part in the 35th General Assembly of the World Medical Association*, 29 October 1983: AAS 76 (1984) 390.

(5) Cf. Declaration *Dignitatis Humanae*, 2.

(6) Pastoral Constitution *Gaudium et Spes*, 22; POPE JOHN PAUL II, Encyclical *Redemptor Hominis*, 8: AAS 71 (1979) 270-272.

(7) Cf. Pastoral Constitution *Gaudium et Spes*, 35.

(8) Pastoral Constitution *Gaudium et Spes*, 15; cf. also POPE PAUL VI, Encyclical *Populorum Progressio*, 20: AAS 59 (1967) 267; POPE JOHN PAUL II, Encyclical *Redemptor Hominis*, 15: AAS 71 (1979) 286-289; Apostolic Exhortation *Familiaris Consortio*, 8: AAS 74 (1982) 89.

(9) POPE JOHN PAUL II, Apostolic Exhortation *Familiaris Consortio*, 11: AAS 74 (1982) 92.

(10) Cf. POPE PAUL VI, Encyclical *Humanae Vitae*, 10: AAS 60 (1968) 487-488.

(11) POPE JOHN PAUL II, *Discourse to the members of the 35th General Assembly of the World Medical Association*, 29 October 1983: AAS 76 (1984) 393.

(12) Cf. POPE JOHN PAUL II, Apostolic Exhortation *Familiaris Consortio*, 11: AAS 74 (1982) 91-92; cf. also Pastoral Constitution *Gaudium et Spes*, 50.

(13) SACRED CONGREGATION FOR THE DOCTRINE OF THE FAITH, *Declaration on Procured Abortion*, 9, AAS 66 (1974) 736-737.

(14) POPE JOHN PAUL II, *Discourse to those taking part in the 35th General Assembly of the World Medical Association*, 29 October 1983: AAS 76 (1984) 390.

(15) POPE JOHN XXIII, Encyclical *Mater et Magistra*, III: AAS 53 (1961) 447.

(16) Pastoral Constitution *Gaudium et Spes*, 24.

(17) Cf. POPE PIUS XII, Encyclical *Humani Generis*: AAS 42 (1950) 575; POPE PAUL VI, *Professio Fidei*: AAS 60 (1968) 436.

(18) POPE JOHN XXIII, Encyclical *Mater et Magistra*, III: AAS 53 (1961) 447; cf. POPE JOHN PAUL II, *Discourse to priests participating in a seminar on "Responsible Procreation"*, 17 September 1983, *Insegnamenti di Giovanni Paolo II*, VI, 2 (1983) 562: "At the origin of each human person there is a creative act of God: no man comes into existence by chance; he is always the result of the creative love of God".

(19) Cf. Pastoral Constitution *Gaudium et Spes*, 24.

(20) Cf. POPE PIUS XII, *Discourse to the Saint Luke Medical-Biological Union*, 12 November 1944: *Discorsi e Radiomessaggi* VI (1944-1945) 191-192.

(21) Cf. Pastoral Constitution *Gaudium et Spes*, 50.

(22) Cf. Pastoral Constitution *Gaudium et Spes*, 51: "When it is a question of harmonizing married love with the responsible transmission of life, the moral character of one's behaviour does not depend only on the good intention and the evaluation of the motives: the objective criteria must be used, criteria drawn from the nature of the human person and human acts, criteria which respect the total meaning of mutual self-giving and human procreation in the context of true love".

(23) Pastoral Constitution *Gaudium et Spes*, 51.

(24) HOLY SEE, *Charter of the Rights of the Family*, 4: L'Osservatore Romano, 25 November 1983.

(25) SACRED CONGREGATION FOR THE DOCTRINE OF THE FAITH, *Declaration on Procured Abortion*, 12-13: AAS 66 (1974) 738.

(26) Cf. POPE PAUL VI, *Discourse to participants in the Twenty-third National Congress of Italian Catholic Jurists*, 9 December 1972: AAS 64 (1972) 777.

(27) The obligation to avoid disproportionate risks involves an authentic respect for human beings and the uprightness of therapeutic intentions. It implies that the doctor "above all ... must carefully evaluate the possible negative consequences which the necessary use of a particular exploratory technique may have upon the unborn child and avoid recourse to diagnostic procedures which do not offer sufficient guarantees of their honest purpose and substantial

harmlessness. And if, as often happens in human choices, a degree of risk must be undertaken, he will take care to assure that it is justified by a truly urgent need for the diagnosis and by the importance of the results that can be achieved by it for the benefit of the unborn child himself" (POPE JOHN PAUL II, *Discourse to Participants in the Pro-Life Movement Congress*, 3 December 1982: *Insegnantenti di Giovanni Paolo II*, V, 3 [1982] 1512). This clarification concerning "proportionate risk" is also to be kept in mind in the following sections of the present Instruction, whenever this term appears.

(28) POPE JOHN PAUL II, *Discourse to the Participants in the 35th General Assembly of the World Medical Association*, 29 October 1983: AAS 76 (1984) 392.

(29) Cf. POPE JOHN PAUL II, *Address to a Meeting of the Pontifical Academy of Sciences*, 23 October 1982: AAS 75 (1983) 37: "I condemn, in the most explicit and formal way, experimental manipulations of the human embryo, since the human being, from conception to death, cannot be exploited for any purpose whatsoever".

(30) HOLY SEE, *Charter of the Rights of the Family*, 4b: *L'Osservatore Romano*, 25 November 1983.

(31) Cf. POPE JOHN PAUL II, *Address to the Participants in the Convention of the Pro-Life Movement*, 3 December 1982: *Insegnamenti di Giovanni Paolo II*, V, 3 (1982) 1511: "Any form of experimentation on the foetus that may damage its integrity or worsen its condition is unacceptable, except in the case of a final effort to save it from death". SACRED CONGREGATION FOR THE DOCTRINE OF THE FAITH, *Declaration on Euthanasia*, 4: AAS 72 (1980) 550: "In the absence of other sufficient remedies, it is permitted, with the patient's consent, to have recourse to the means provided by the

most advanced medical techniques, even if these means are still at the experimental stage and are not without a certain risk".

(32) No one, before coming into existence, can claim a subjective right to begin to exist; nevertheless, it is legitimate to affirm the right of the child to have a fully human origin through conception in conformity with the personal nature of the human being. Life is a gift that must be bestowed in a manner worthy both of the subject receiving it and of the subjects transmitting it. This statement is to be borne in mind also for what will be explained concerning artificial human procreation.

(33) Cf. POPE JOHN PAUL II, *Discourse to those taking part in the 35th General Assembly of the World Medical Association*, 29 October 1983: AAS 76 (1984) 391.

(34) Cf. Pastoral Constitution on the Church in the Modern world, *Gaudium et Spes*, 50.

(35) Cf. POPE JOHN PAUL II, Apostolic Exhortation *Familiaris Consortio*, 14: AAS 74 (1982) 96.

(36) Cf. POPE PIUS XII, *Discourse to those taking part in the 4th International Congress of Catholic Doctors*, 29 September 1949: AAS 41 (1949) 559. According to the plan of the Creator, "A man leaves his father and his mother and cleaves to his wife, and they become one flesh" (*Gen* 2:24). The unity of marriage, bound to the order of creation, is a truth accessible to natural reason. The Church's Tradition and Magisterium frequently make reference to the Book of Genesis, both directly and through the passages of the New Testament that refer to it: *Mt* 19: 4-6; *Mk*: 10:5-8; *Eph* 5: 31. Cf. ATHENAGORAS, *Legatio pro christianis*, 33: PG 6, 965-967; ST CHRYSOSTOM, *In Matthaeum homiliae*, LXII, 19, 1: PG 58 597; ST LEO

THE GREAT, *Epist. ad Rusticum*, 4: PL 54, 1204; INNOCENT III, *Epist. Gaudemus in Domino*: DS 778; COUNCIL OF LYONS II, IV *Session*: DS 860; COUNCIL OF TRENT, XXIV , *Session*: DS 1798. 1802; POPE LEO XIII, Encyclical *Arcanum Divinae Sapientiae*: ASS 12 (1879/80) 388-391; POPE PIUS XI, Encyclical *Casti Connubii*: AAS 22 (1930) 546-547; SECOND VATICAN COUNCIL, *Gaudium et Spes*, 48; POPE JOHN PAUL II, Apostolic Exhortation *Familiaris Consortio*, 19: AAS 74 (1982) 101-102; *Code of Canon Law*, Can.1056.

(37) Cf. POPE PIUS XII, *Discourse to those taking part in the 4th International Congress of Catholic Doctors*, 29 September 1949: AAS 41 (1949) 560; *Discourse to those taking part in the Congress of the Italian Catholic Union of Midwives*, 29 October 1951: AAS 43 (1951) 850; *Code of Canon Law*, Can. 1134.

(38) POPE PAUL VI, Encyclical Letter *Humanae Vitae*, 12: AAS 60 (1968) 488-489.

(39) *Loc. cit., ibid.*, 489.

(40) POPE PIUS XII, *Discourse to those taking part in the Second Naples World Congress on Fertility and Human Sterility*, 19 May 1956: AAS 48 (1956) 470.

(41) *Code of Canon Law*, Can. 1061. According to this Canon, the conjugal act is that by which the marriage is consummated if the couple "have performed (it) between themselves in a human manner".

(42) Cf. Pastoral Constitution *Gaudium et Spes*, 14.

(43) Cf. POPE JOHN PAUL II, *General Audience on 16 January 1980*: *Insegnamenti di Giovanni Paolo II*, III, 1 (1980) 148-152.

(44) POPE JOHN PAUL II, *Discourse to those taking part in the 35th General Assembly of the World Medical Association*, 29 October 1983: AAS 76 (1984) 393.

(45) Cf. Pastoral Constitution *Gaudium et Spes*, 51.

(46) Cf. Pastoral Constitution *Gaudium et Spes*, 50.

(47) Cf. POPE PIUS XII, *Discourse to those taking part in the 4th International Congress of Catholic Doctors*, 29 September 1949: AAS 41 (1949) 560: "It would be erroneous ... to think that the possibility of resorting to this means (artificial fertilization) might render valid a marriage between persons unable to contract it because of the *impedimentum impotentiae*".

(48) A similar question was dealt with by POPE PAUL VI, Encyclical *Humanae Vitae*, 14: AAS 60 (1968) 490-491.

(49) Cf. *supra*: I, 1 ff.

(50) POPE JOHN PAUL II, Apostolic Exhortation *Familiaris Consortio*. 14: AAS 74 (1982) 96.

(51) Cf. *Response of the Holy Office*, 17 March 1897: DS 3323; POPE PIUS XII, *Discourse to those taking part in the 4th International Congress of Catholic Doctors*, 29 September 1949: AAS 41 (1949) 560; *Discourse to the Italian Catholic Union of Midwives*, 29 October 1951: AAS 43 (1951) 850; *Discourse to those taking part in the Second Naples World Congress on Fertility and Human Sterility*, 19 May 1956: AAS 48 (1956) 471-473; *Discourse to those taking part in the 7th International Congress of the International Society of Haematology*, 12 September 1958: AAS 50 (1958) 733; POPE JOHN XXIII, Encyclical *Mater et Magistra*, III: AAS 53 (1961) 447.

(52) POPE PIUS XII, *Discourse to the Italian Catholic Union of Midwives*, 29 October 1951: AAS 43 (1951) 850.

(53) POPE PIUS XII, *Discourse to those taking part in the 4th International Congress of Catholic Doctors*, 29 September 1949: AAS 41 (1949) 560.

(54) SACRED CONGREGATION FOR THE DOCTRINE OF THE FAITH, *Declaration on Certain Questions Concerning Sexual ethics*, 9: AAS 68 (1976) 86, which quotes the Pastoral Constitution *Gaudium et Spes*, 51. Cf. *Decree of the Holy Office*, 2 August 1929: AAS 21 (1929) 490; POPE PIUS XII, *Discourse to those taking part in the 26th Congress of the Italian Society of Urology*, 8 October 1953: AAS 45 (1953) 678.

(55) Cf. POPE JOHN XXIII, Encyclical *Mater et Magistra*, III: AAS 53 (1961) 447.

(56) Cf. POPE PIUS XII, *Discourse to those taking part in the 4th International Congress of Catholic Doctors*, 29 September 1949: AAS 41 (1949), 560.

(57) Cf. POPE PIUS XII, *Discourse to the taking part in the Second Naples World Congress on Fertility and Human Sterility*, 19 May 1956: AAS 48 (1956) 471-473.

(58) Pastoral Constitution *Gaudium et Spes*, 50.

(59) POPE JOHN PAUL II, Apostolic Exhortation *Familiaris Consortio*, 14: AAS 74 (1982) 97.

(60) Cf. Declaration *Dignitatis Humanae*, 7.

www.ingramcontent.com/pod-product-compliance
Lightning Source LLC
Chambersburg PA
CBHW060431050426
42449CB00009B/2243